Addition & Subtra
Age 6-7

Alison Oliver

In a strange place, not too far from here, lives a scare of monsters.

A 'scare' is what some people call a group of monsters, but these monsters are really very friendly once you get to know them.

They are a curious bunch – they look very unusual, but they are quite like you and me, and they love learning new things and having fun.

In this book you will go on a learning journey with the monsters and you are sure to have lots of fun along the way.

Do not forget to visit our website to find out more about all the monsters and to send us photos of you in your monster mask or the monsters that you draw and make!

Contents

Monster Trios

Webber is making a monster potion using these ingredients:

9 Moodle teeth

2 cups of leech juice

8 slimy eels

6 Leckie hairs

The first instruction is to **add together** the Moodle teeth and the Leckie hairs.

What is 9 Moodle teeth **add** 6 Leckie hairs?
Count on a number line.

| 6 | 7 | 8 | 9 | 10 | 11 | 12 | 13 | 14 | 15 | 16 | 17 |

$9 + 6 = 15$

This is a **trio**.
A trio is a set of three numbers that make **addition** and **subtraction** families.

$9 + 6 = 15$ so that means $6 + 9 = 15$

We can also count back (subtract).
$15 - 6 = 9$ and $15 - 9 = 6$

1   Use the number line to write the answer for each of these.

| 0 | 1 | 2 | 3 | 4 | 5 | 6 | 7 | 8 | 9 | 10 | 11 | 12 | 13 | 14 | 15 | 16 | 17 | 18 | 19 | 20 |

a $8 + 5 =$ ☐

b $7 + 8 =$ ☐

c $2 + 17 =$ ☐

d $12 + 6 =$ ☐

e $11 + 0 =$ ☐

f $13 - 4 =$ ☐

g $15 - 3 =$ ☐

h $18 - 7 =$ ☐

i $20 - 8 =$ ☐

2 Write four number sentences for each set of pictures.

a

7	+		=	
13	–	6	=	
6	+		=	
	–	7	=	

b

	+		=	
15	–		=	
	+		=	
15	–		=	

3 Write the missing numbers by adding and subtracting.

14 → + 6 → 20 → – 9 → → – 5 → → + 7 → → – 7 → → + 8 → → + 2 → → – 3 →

Fun Zone!

Turn these shapes into monsters.

Well done! You can now find and colour **Shape 1** on the Monster Match page!

The Inverse

Adding and taking away are opposites.
We call them the **inverse** calculations.

Webber is sharing some monster bites
with Grandpa.
There are 17 monster bites.
They eat 8 of them.
How many are left?

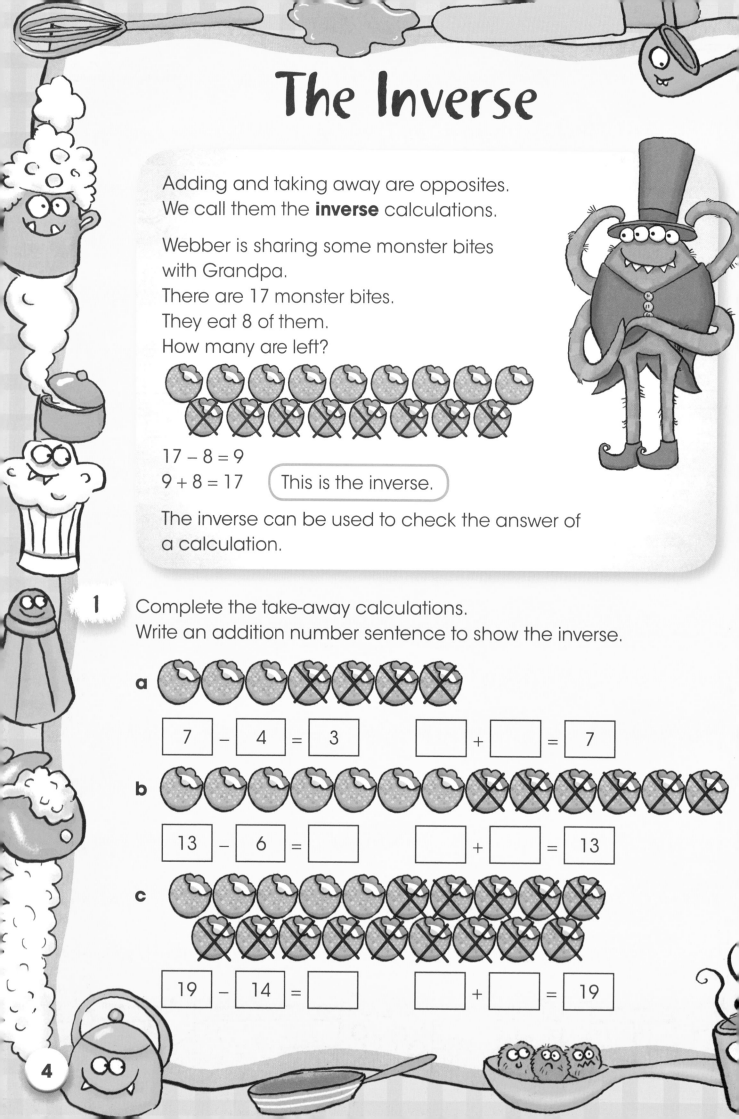

$17 - 8 = 9$
$9 + 8 = 17$ ⟮ This is the inverse. ⟯

The inverse can be used to check the answer of
a calculation.

1 Complete the take-away calculations.
Write an addition number sentence to show the inverse.

a

| 7 | – | 4 | = | 3 | | | + | | = | 7 |

b

| 13 | – | 6 | = | | | | + | | = | 13 |

c

| 19 | – | 14 | = | | | | + | | = | 19 |

4

2 Answer these.
Use the number line to check your calculations.

$\boxed{8}\ \boxed{9}\ \boxed{10}\ \boxed{11}\ \boxed{12}\ \boxed{13}\ \boxed{14}\ \boxed{15}\ \boxed{16}\ \boxed{17}\ \boxed{18}\ \boxed{19}\ \boxed{20}$

a $\boxed{14} + \boxed{5} = \boxed{}$ $\boxed{19} - \boxed{} = \boxed{14}$

b $\boxed{12} + \boxed{3} = \boxed{}$ $\boxed{15} - \boxed{} = \boxed{12}$

c $\boxed{9} + \boxed{8} = \boxed{}$ $\boxed{} - \boxed{8} = \boxed{}$

d $\boxed{11} + \boxed{9} = \boxed{}$ $\boxed{} - \boxed{11} = \boxed{}$

3 Colour the matching inverse calculations the same colour.
The first one has been done for you.

| 10 – 4 | 19 – 8 | 8 + 9 | 16 – 3 | 6 + 5 |

| 11 + 8 | 11 – 5 | 13 + 3 | 17 – 8 | 6 + 4 |

Fun Zone!

Time to make some footprint monsters.

Well done! You can now find **Shape 2** on the Monster Match page and colour it in!

Footprint Monsters

You will need a piece of card, coloured paint, paper, crayons, glue and some old newspaper to cover the floor.

Ask an adult to help when needed.

1 Paint the underside of your foot and stamp it on the card.

2 Allow the paint to dry.

3 Draw a set of eyes on the paper and glue them onto your monster.

4 Use crayons and coloured paint to decorate your monster.

Adding in any Order

Webber is taking Zak for a walk.
He sees that Tizz, Fizz and Poggo
are picking daisies.
They are trying to add up how
many daisies they have altogether.
When we add numbers we find
the **total** (or **sum**).

Here are three ways to find the sum of three numbers.

(1) Tizz has 3 daisies, Fizz has 9 and Poggo has 5.
We can **add in any order**.
Start with the largest number.
9 + 5 + 3 = 17

(2) Look for **number bonds** that make 10.
Then add on the third number.
6 + 4 + 7   **= 10 + 7**   **= 17**

(3) Watch out for two numbers in calculations that are
the same.
These are called **doubles**.
Double 8 = 16   8 + 8 = 16
Doubles can be used to help total numbers that are
near doubles.
8 + 8 = 16   8 + 9 is 1 more than 16   8 + 9 = 17

1 Re-order these numbers, starting with the largest.
Then find the sum.

a 2 + 4 + 9   ☐ + ☐ + ☐ = ☐

b 8 + 4 + 5   ☐ + ☐ + ☐ = ☐

c 5 + 0 + 7   ☐ + ☐ + ☐ = ☐

2 Use your number bonds to answer these.
The first one has been done for you.

a $9 + 7 + 1$ = | 9 | + | 1 | + | 7 | = | 10 | + | 7 | = | 17 |

b $2 + 6 + 8$ = | ☐ | + | ☐ | + | ☐ | = | ☐ | + | 6 | = | ☐ |

c $3 + 5 + 7$ = | ☐ | + | ☐ | + | ☐ | = | ☐ | + | ☐ | = | ☐ |

3 Use doubles to add these three numbers.

a ☐ + ☐ + ☐ = ☐

b ☐ + ☐ + ☐ = ☐

c ☐ + ☐ + ☐ = ☐

d ☐ + ☐ + ☐ = ☐

Fun Zone!

Make new words using letters in the word 'monster'.

That is a lot of words! You can now find and colour **Shape 3** on the Monster Match page!

M O N S T E R

Here are some examples:
SON
REST
STONE

Monster Number Lines

Webber is helping Poggo with his maths homework.
It is very tricky!
Webber shows him that number lines are a great way to add and subtract when you start working with bigger numbers.

You can count forwards in 1s.

$27 + 4 = 31$

25 26 27 28 29 30 31 32 33 34 35 36 37 38

You can count backwards in 1s.

$34 - 3 = 31$

25 26 27 28 29 30 31 32 33 34 35 36 37 38

Remember, adding makes a bigger number.
Subtracting means taking away, leaving a smaller number.

1 Write the next three numbers in each sequence.

a	28	27	26			
b	69	70	71			
c	57	58	59			
d	22	21	20			
e	87	88	89			

2 Use the number lines to write the number sentences.
The first one has been done for you.

a 37 38 39 40 41 42 43 44 45 46 47 | 44 | – | 5 | = | 39 |

b 75 76 77 78 79 80 81 82 83 84 85 | | + | | = | |

c 30 31 32 33 34 35 36 37 38 39 40 | | – | | = | |

d 78 79 80 81 82 83 84 85 86 87 88 | | – | | = | |

3 Answer these.
Use the number lines to help you.

a 37 + 8 = [ ]   34 35 36 37 38 39 40 41 42 43 44 45 46 47 48

b 23 + 9 = [ ]   20 21 22 23 24 25 26 27 28 29 30 31 32 33 34

c 87 + 3 = [ ]   82 83 84 85 86 87 88 89 90 91 92 93 94 95 96

d 57 – 7 = [ ]   45 46 47 48 49 50 51 52 53 54 55 56 57 58 59

Fun Zone!

Help the monsters
make their way
through the maze
from start to finish.

Congratulations! You
can now find and
colour **Shape 4** on the
Monster Match page!

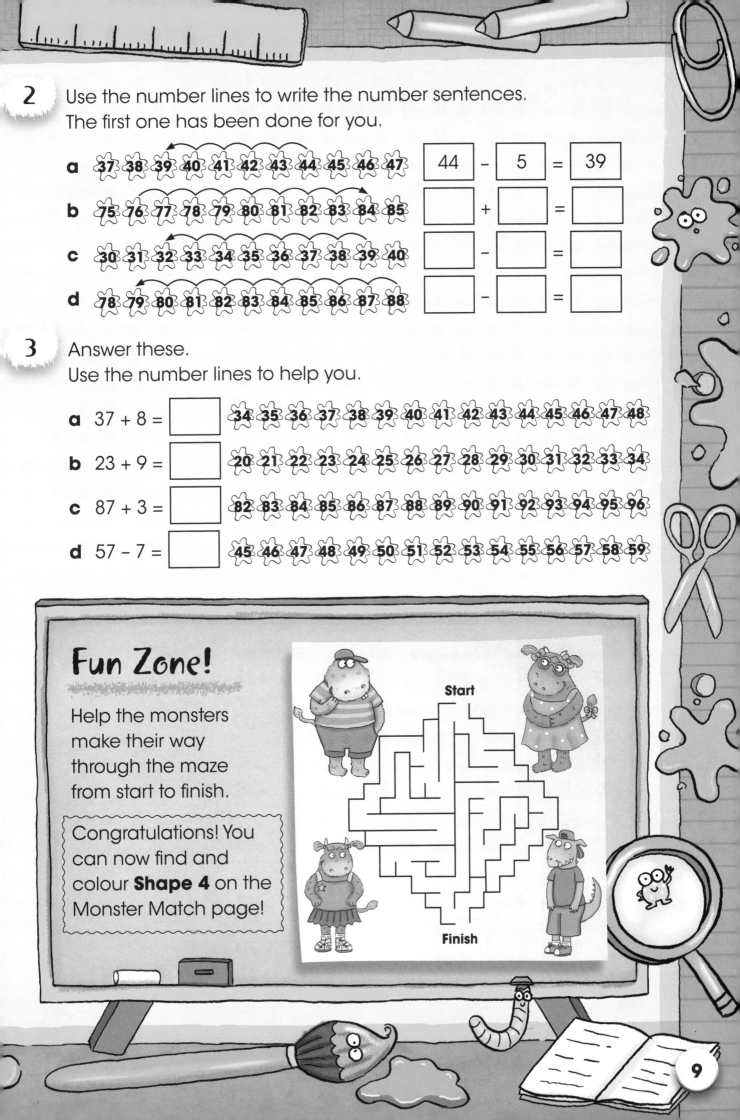

Start

Finish

Monster Multiples of 10

Poggo and I have found some 10-legged hairy caterpillars in the cave.

We wonder how many legs the caterpillars have altogether?

You already know how to use **number bonds**. So, now it is easy to add **multiples of 10**. Look at the pattern in these.

$$3 + 6 = 9 \qquad\qquad 4 + 7 = 11$$
$$30 + 60 = 90 \qquad\qquad 40 + 70 = 110$$

These are all multiples of 10.

(10)(20)(30)(40)(50)(60)(70)(80)(90)(100)(110)(120)(130)

1 Write the answers to complete these number sentences.

a $4 + 3 = \boxed{}$

$40 + 30 = \boxed{}$

d $5 + 3 = \boxed{}$

$50 + 30 = \boxed{}$

b $9 + 4 = \boxed{}$

$90 + 40 = \boxed{}$

e $7 + 5 = \boxed{}$

$70 + 50 = \boxed{}$

c $6 + 2 = \boxed{}$

$60 + 20 = \boxed{}$

f $8 + 7 = \boxed{}$

$80 + 70 = \boxed{}$

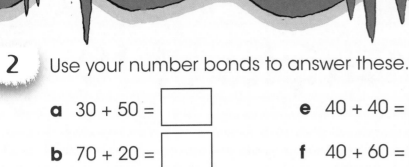

2 Use your number bonds to answer these.

a 30 + 50 = ☐ **e** 40 + 40 = ☐

b 70 + 20 = ☐ **f** 40 + 60 = ☐

c 60 + 50 = ☐ **g** 80 + 60 = ☐

d 90 + 40 = ☐ **h** 70 + 90 = ☐

3 Draw a line to match the incomplete number sentence on the caterpillar with its answer on the rock.

 90 + 70 70 + 60 80 + 70

130 90 150 140 110 160

 40 + 50 30 + 80 60 + 80

Fun Zone!

Try to imagine your own monsters. List your ideas on a separate piece of paper.

Great monster ideas! You can now find and colour **Shape 5** on the Monster Match page!

Write down a description for your own monster.

- What body parts will you give it?
- How will its voice sound?
- What will it smell like?
- What colour will it be?
- How many eyes will it have?

More Multiples of 10

Grandpa is fixing the slide in the park.
He has a big pile of different coloured screws.
He only needs blue screws to fix the slide.
He has started to sort the blue screws
into a separate pile.
He asks Poggo to add another
10 blue screws from the big pile
to the blue pile.

When we **add or subtract 10** to or from
a small number, the **units stay the same**.
Only the digit for the tens changes.
Can you see the pattern?

3 + 10 = **1**3	35 – 10 = **2**5
13 + 10 = **2**3	25 – 10 = **1**5
23 + 10 = **3**3	15 – 10 = 5

1 Add 10 to these.
The first one has been done for you.

a 15 + 10 = ☐ 25

c 37 + 10 = ☐

b 76 + 10 = ☐

d 43 + 10 = ☐

2 Add a multiple of 10 to these.
Remember, 30 is three lots of 10.
The first one has been done for you.

a 24 + 20 = ☐ 44

c 40 + 40 = ☐

b 55 + 30 = ☐

d 12 + 40 = ☐

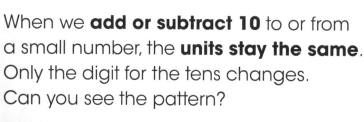

3 Take away 10 from these.
The first one has been done for you.

a 32 – 10 = 22

b 17 – 10 = ☐

c 44 – 10 = ☐

d 36 – 10 = ☐

4 Now try taking away a multiple of 10.
The first one has been done for you.

a 22 – 20 = 2

b 99 – 40 = ☐

c 58 – 30 = ☐

d 61 – 20 = ☐

5 Draw lines to match the calculation to the correct answer.

| 4 + 30 | | 26 + 40 | | 52 – 20 |

7   34   19   32   66   100

| 17 – 10 | | 80 + 20 | | 39 – 20 |

Fun Zone!

Make your own peg monster.

Watch that monster does not bite you! You can now find and colour **Shape 6** on the Monster Match page!

Snapping Peg Monsters

You will need coloured paper, white paper, clothes pegs, crayons and glue.
Ask an adult to help when needed.

1 Draw your monster in two separate halves on the coloured paper. The two halves will be stuck on the pegs. Cut them out.

2 Draw some teeth on white paper and cut them out.

3 Glue the teeth onto the back of your monster, so that the sharp teeth poke out.

4 Glue the halves of your monster to the top and bottom of the peg.

Monster Challenge 1

1 Complete these addition squares.
For each square, add the top number to the number on the
far left.

a

+	8	7	6
4			
7		14	
9			

b

+	13	12	8
7			
6			
5	18		

2 Here are sets of numbers.
Write two inverse number sentences for each set.

a 5 6 11   ☐ + ☐ = ☐   ☐ − ☐ = ☐

b 10 4 14   ☐ − ☐ = ☐   ☐ + ☐ = ☐

c 2 7 5   ☐ + ☐ = ☐   ☐ − ☐ = ☐

d 9 4 5   ☐ − ☐ = ☐   ☐ + ☐ = ☐

3 Write a number sentence that adds two small numbers to the start
number so that you land on the total.
Use the number line to help you.
The first one has been done for you.

0 1 2 3 4 5 6 7 8 9 10 11 12 13 14 15 16 17 18 19 20

a Start at 4 and land on 15 | 4 + 4 + 7 = 15 |

b Start at 7 and land on 12

c Start at 2 and land on 13

d Start at 7 and land on 20

4 Use the monsterbots to find the missing numbers.

a

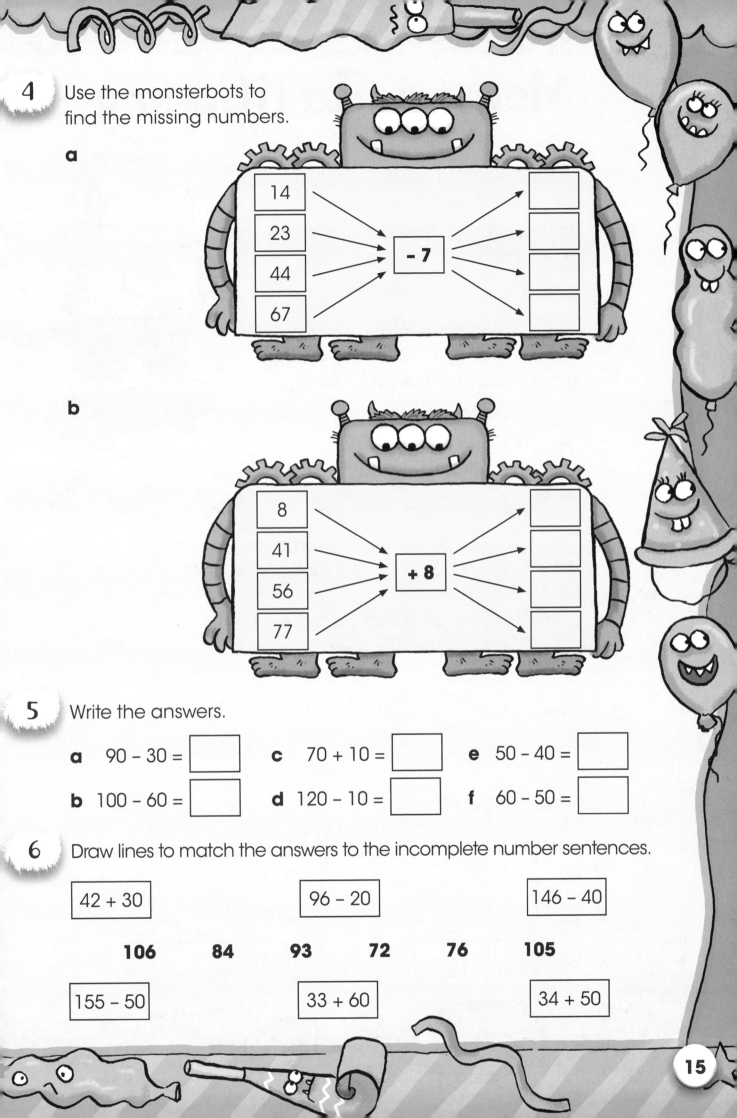

| 14 |
| 23 |
| 44 |
| 67 |

−7

b

| 8 |
| 41 |
| 56 |
| 77 |

+8

5 Write the answers.

a 90 − 30 = ☐   **c** 70 + 10 = ☐   **e** 50 − 40 = ☐

b 100 − 60 = ☐   **d** 120 − 10 = ☐   **f** 60 − 50 = ☐

6 Draw lines to match the answers to the incomplete number sentences.

| 42 + 30 | | 96 − 20 | | 146 − 40 |

106   **84**   **93**   **72**   **76**   **105**

| 155 − 50 | | 33 + 60 | | 34 + 50 |

Monster Partitioning

Poggo and Fizz use Mum's beads to split tens and ones in different ways.
Think about the **tens** and **units** in a number separately.

$37 = 30 + 7 \qquad 50 = 50 + 0 \qquad 11 = 10 + 1$

This is called **partitioning**.
If a calculation looks tricky, **partition** the numbers and use your facts about multiples of 10!

$37 + 21 = 30 + 7 + 20 + 1$

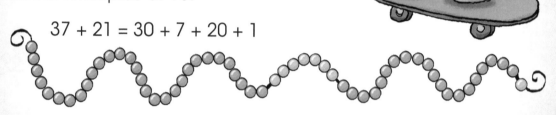

$= 50 + 7 + 1$ **Combine** the tens
$= 58$ **Add** on the units

Sometimes the units make a group of ten.
$$26 + 27 = 20 + 20 + 6 + 7$$
$$= 40 + 13$$
$$= 40 + 10 + 3$$
$$= 53$$

1   Partition these numbers.
The first one has been done for you.

a   45 = ☐ 40 ☐ + ☐ 5 ☐   **d**   15 = ☐ + ☐

b   23 = ☐ + ☐   **e**   70 = ☐ + ☐

c   98 = ☐ + ☐   **f**   39 = ☐ + ☐

2 Partition the numbers.
Complete the answers.

a 27 + 31=

$\boxed{20}$ + $\boxed{30}$ + $\boxed{7}$ + $\boxed{1}$ = $\boxed{}$ + $\boxed{7}$ + $\boxed{1}$ = $\boxed{}$

b 42 + 34 =

$\boxed{40}$ + $\boxed{}$ + $\boxed{2}$ + $\boxed{}$ = $\boxed{70}$ + $\boxed{}$ + $\boxed{}$ = $\boxed{}$

3 Colour the frogs to match the colour of the nets.

41 + 15   31 + 34   24 + 32   21 + 52

56   **65**   **73**

42 + 31   21 + 44   29 + 27   19 + 37

Fun Zone!

Monster colour by numbers. The key shows you which colour to use in each area.

Well done! You can now find and colour **Shape 9** on the Monster Match page!

Key
- 1
- 2
- 3
- 4

More Monster Partitioning

Webber walks from Monsterville to the wild wood in 56 steps.
On the way back, he finds a shortcut.
The shortcut is only 27 steps.
He wants to know the difference between the two routes.

Partitioning can be useful when we subtract too.
Use the number line below to help.

$56 - 27 = ?$

Remember $27 = 20 + 7$.
First jump back 20, then jump back 7.

29 30 31 32 33 34 35 36 37 38 39 40 41 42 43 44 45 46 47 48 49 50 51 52 53 54 55 56

So, $56 - 27 = 29$.

1 Use the number lines to answer these.

a $48 - 16 = \boxed{}$

32 33 34 35 36 37 38 39 40 41 42 43 44 45 46 47 48

b $56 - 13 = \boxed{}$

40 41 42 43 44 45 46 47 48 49 50 51 52 53 54 55 56

c $75 - 14 = \boxed{}$

59 60 61 62 63 64 65 66 67 68 69 70 71 72 73 74 75

2 Write the missing answers on the cobbles.

a Start on 90 and work backwards, minus 15 each jump.

(90) ( ) (60) ( ) (30) ( ) (0)

b Start on 77 and find the number that is 11 less.

( ) (11) ( ) (33) ( ) ( ) (66) (77)

3 Answer these.
Colour the answer that is the odd one out in each row.

a 43 – 21 = ☐   54 – 32 = ☐   62 – 40 = ☐   36 – 15 = ☐

b 74 – 21 = ☐   85 – 33 = ☐   96 – 43 = ☐   87 – 34 = ☐

c 56 – 12 = ☐   65 – 31 = ☐   77 – 43 = ☐   58 – 24 = ☐

Fun Zone!

Time to make a monster face!

Scary! You can now find and colour **Shape 8** on the Monster Match page!

Paper Plate Monster

You will need a paper plate, crayons, scissors, glue, paper, coloured paint and paintbrushes.

Ask an adult to help when needed.

1 Paint your plate and leave to dry.
2 Cut out some white paper triangles for teeth and ears.
3 Glue the teeth and ears onto your plate.
4 Make some eyes using paper and crayons, cut them out and stick them on.
5 Decorate your monster face with crayons.

Missing Monster Numbers

Webber is feeding Leckie.
Oh no!
There were 50 pet biscuits
but now there are only 26.
Can you work out how many
Leckie has eaten?

$50 - ? = 26$
You can count back to find the difference!

24 25 26 27 28 29 30 31 32 33 34 35 36 37 38 39 40 41 42 43 44 45 46 47 48 49 50

Add up the jumps.
$10 + 10 + 4 = 24$
So, $50 - 24 = 26$
$? = 24$

1 Find the missing numbers.
Use the number lines to help you.

a $24 - \boxed{} = 12$

12 13 14 15 16 17 18 19 20 21 22 23 24 25

b $35 - \boxed{} = 18$

18 19 20 21 22 23 24 25 26 27 28 29 30 31 32 33 34 35

c $52 - \boxed{} = 38$

36 37 38 39 40 41 42 43 44 45 46 47 48 49 50 51 52

2 Fill in the missing numbers on the boxes of pet biscuits.

Biscuits = 60
( ) + (35)

Biscuits = 77
( ) + (40)

Biscuits = 23
(14) + ( )

Biscuits = 94
(87) + ( )

Biscuits = 27
( ) + (18)

Biscuits = 55
(42) + ( )

Biscuits = 31
( ) + (14)

3 Answer these.
Colour each box of pet biscuits using the key.

> 15 green   16 blue   17 yellow   18 red

a $7 + 8 = \boxed{}$

b $98 - \boxed{} = 80$

c $\boxed{} + 15 = 32$

d $30 - \boxed{} = 13$

e $\boxed{} - 8 = 8$

f $76 - 58 = \boxed{}$

g $5 + 12 = \boxed{}$

h $7 + 9 = \boxed{}$

i $55 - \boxed{} = 40$

Fun Zone!

Find each group of numbers in order in the number search. Circle each group of numbers in the grid.

12	13	14	15	16
50	51	52	53	54
38	39	40	41	42
82	83	84	85	86
68	69	70	71	72

50	12	13	14	15	16	42
20	51	4	34	87	22	41
82	18	52	65	11	35	40
83	44	1	53	67	94	39
84	29	88	3	54	32	38
85	72	71	70	69	68	24
86	99	6	5	7	8	9

Well done! You can now find and colour **Shape 7** on the Monster Match page!

Monster Number Puzzles

I love puzzle books!
Sometimes numbers are written with words instead of digits.
It is a good way to practise your spelling.
Watch out for the tricky spellings!

Here are some symbols and their meanings:

< means less than   **> means greater than**

1 < 2   and   8 > 4
one is less   and   eight is greater
than two   than four

1 Write the answers to these subtractions in words.
The first one has been done for you.
What is the secret number in the shaded boxes?

a 21 – 18   c 19 – 13   e 20 – 8   g 11 – 2
b 32 – 24   d 49 – 45   f 15 – 8   f 100 – 99

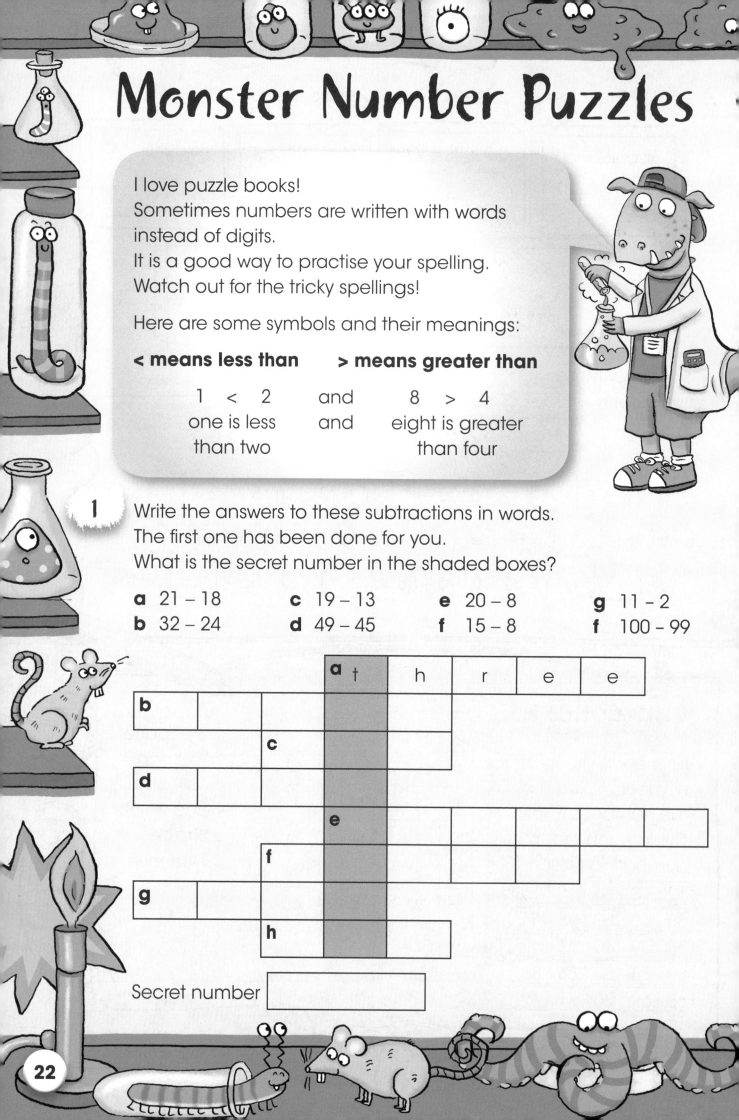

a | t | | h | r | e | e

b

c

d

e

f

g

h

Secret number

22

2 Use the < or > symbol so that these sentences make sense.

a Five ☐ ten

c Ninety-nine ☐ thirty-six

b Eleven ☐ seven

d Fifty-one ☐ twelve

3 Write the number sentence for each problem.

a There are usually thirty monsters in Class 3 but today sixteen are on a school trip.
How many monsters are in Class 3 today?

_____

b There are twenty-eight monsters in Class 4.
How many monsters are there in Class 3 and Class 4 altogether normally?

_____

c Usually in Class 3, eleven monsters do not wear glasses.
How many monsters in Class 3 usually wear glasses?

_____

Fun Zone!

Find all the monster noise words in this word search.

Well done! You can now find and colour **Shape 10** on the Monster Match page!

GROWL
SCREAM
HISS
ROAR
SNARL
GRUNT

Q	G	R	U	N	T	L
C	S	B	M	M	I	W
R	E	N	O	A	J	O
Q	O	R	A	E	Y	R
V	A	A	N	R	E	G
T	O	X	R	C	L	U
P	H	I	S	S	L	W

Monster Maths in Columns

I have found 35 black bugs and
49 red beetles in the cave.
How many insects do I have altogether?
35 + 49 = 84

Here is another way to write it, using **columns**.
Be careful to write **units under units** and
tens under tens.
Using squared paper can help you line
the numbers up.

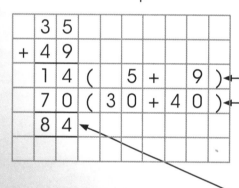

	3	5
+	4	9
	1	4	(5 + 9) ← Add up the **units**.
	7	0	(30 + 40) ← Add up the **tens**.
	8	4

Add the columns,
(4 + 0 = 4 and
10 + 70 = 80).
Here is your total.

1 Answer these.

a

	4	3
+	2	6
			(3 + 6)
			(40 + 20)

c

	6	7
+	3	1
			(7 +)
			(+)

b

	5	1
+	3	6
			(1 + 6)
			(+ 30)

d

	6	4
+	2	5
			(+)
			(+)

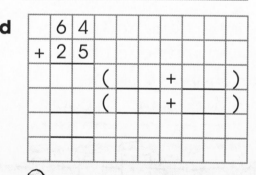

2 Write in the missing numbers.

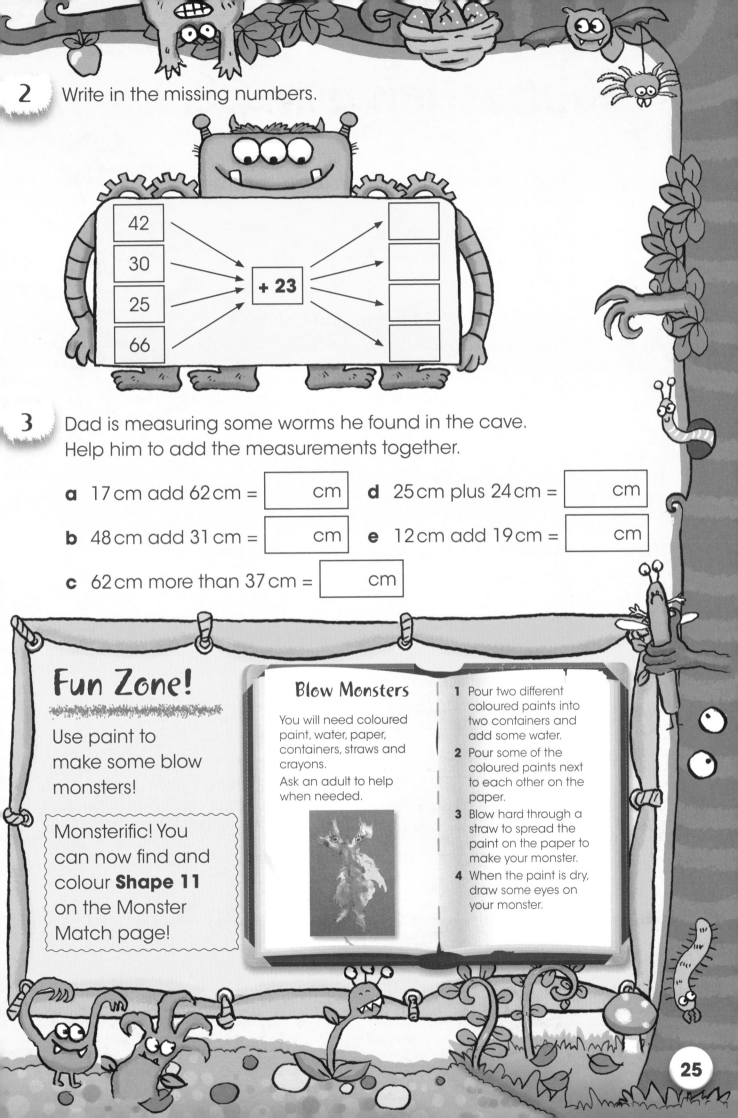

42
30
25
66

+ 23

3 Dad is measuring some worms he found in the cave.
Help him to add the measurements together.

a 17 cm add 62 cm = ☐ cm   **d** 25 cm plus 24 cm = ☐ cm

b 48 cm add 31 cm = ☐ cm   **e** 12 cm add 19 cm = ☐ cm

c 62 cm more than 37 cm = ☐ cm

Fun Zone!

Use paint to make some blow monsters!

Monsterific! You can now find and colour **Shape 11** on the Monster Match page!

Blow Monsters

You will need coloured paint, water, paper, containers, straws and crayons.
Ask an adult to help when needed.

1 Pour two different coloured paints into two containers and add some water.
2 Pour some of the coloured paints next to each other on the paper.
3 Blow hard through a straw to spread the paint on the paper to make your monster.
4 When the paint is dry, draw some eyes on your monster.

Subtraction using Columns

Poggo is throwing a ball for Zak to catch.
Poggo throws the ball 69 times.
Zak drops it 37 times.
How many times does Zak
catch the ball?

You can use **partitioning** and **columns**
for subtraction too.

$69 - 37 = ?$

$69 = 60 + 9$ $37 = 30 + 7$

So

	6	0	+	9
−	3	0	+	7
	3	0	+	2

←——($9 - 7 = 2$ and $60 - 30 = 30$)

Recombine $30 + 2$, to get the final number.
The answer is $69 - 37 = 32$.

1 Try these.
The first one has been done for you.

a $79 - 56 =$

	7	0	+	9		
−	5	0	+	6		
	2	0	+	3	= 2	3

b $47 - 36 =$

c $98 - 23 =$

26

2 Write in the
missing numbers.

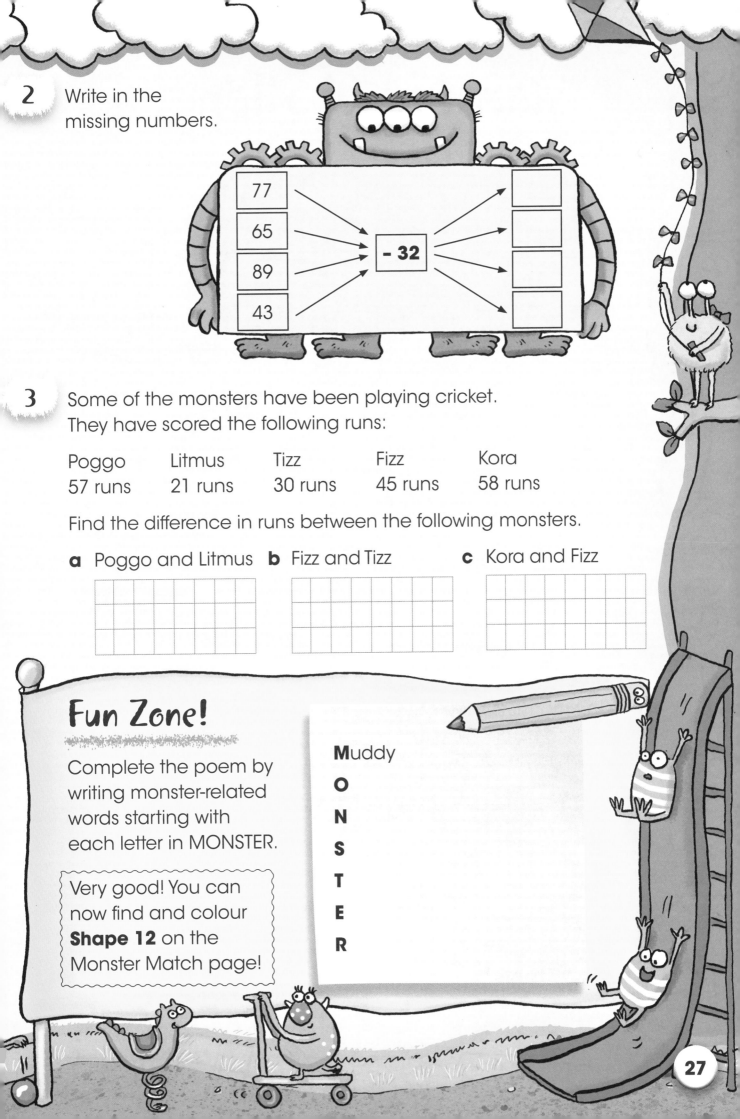

| 77 |
| 65 |
| 89 |
| 43 |

– 32

3 Some of the monsters have been playing cricket.
They have scored the following runs:

Poggo	Litmus	Tizz	Fizz	Kora
57 runs	21 runs	30 runs	45 runs	58 runs

Find the difference in runs between the following monsters.

a Poggo and Litmus   **b** Fizz and Tizz   **c** Kora and Fizz

Fun Zone!

Complete the poem by
writing monster-related
words starting with
each letter in MONSTER.

Very good! You can
now find and colour
Shape 12 on the
Monster Match page!

Muddy
O
N
S
T
E
R

Monster Challenge 2

1 Use your partitioning skills to answer these.

a 52 + 41 =

[ ] + [ ] + [ ] + [ ] = [ ] + [ ] + [ ] = [ ]

b 24 + 73 =

[ ] + [ ] + [ ] + [ ] = [ ] + [ ] + [ ] = [ ]

2 Colour the skateboard with the correct answer.

a 51 – 14   36   37   38

b 95 – 16   79   77   78

c 26 – 19   6   8   7

d 87 – 18   68   69   67

3 The monsters have scored some runs in cricket.
Draw a number line to work out how many more runs each monster needs to score 100.

a 73   73 74 75 76 77 78 79 80 81 82 83 84 85 86 87 88 89 90 91 92 93 94 95 96 97 98 99 **100**

b 81

c 62

d 59

28

4 Rearrange the letters to spell a number.
Draw a line to join each number to the correct digit.
The first one has been done for you.

owt | two | 60

veleen | | 8

hirtty | | 2

tixys | | 30

higet | | 11

5 Use your columns to solve these.

a

```
    3 6
+   2 3
      (     +     )
      (     +     )
```

b

```
    5 7
+   4 1
      (     +     )
      (     +     )
```

6 Work out the change.

a You have 100p.
How much change do you get if you spend 85p?   [ ] p

b You have 38p.
How much change do you get if you spend 18p?   [ ] p

c You have 45p.
How much change do you get if you spend 21p?   [ ] p

d You have 77p.
How much change do you get if you spend 43p?   [ ] p

I knew you could do it!
You have made it to the end of the book.
You are a magnificent monster!

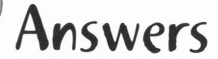

Answers

Page 2

1 **a** 8 + 5 = 13 **f** 13 – 4 = 9
b 7 + 8 = 15 **g** 15 – 3 = 12
c 2 + 17 = 19 **h** 18 – 7 = 11
d 12 + 6 = 18 **i** 20 – 8 = 12
e 11 + 0 = 11

Page 3

2 **a** 7 + 6 = 13 **b** 5 + 10 = 15
13 – 6 = 7 15 – 5 = 10
6 + 7 = 13 10 + 5 = 15
13 – 7 = 6 15 – 10 = 5

3 14 + 6 = 20; 20 – 9 = 11; 11 – 5 = 6; 6 + 7 = 13; 13 – 7 = 6; 6 + 8 = 14; 14 + 2 = 16; 16 – 3 = 13

Page 4

1 **a** 7 – 4 = 3 3 + 4 = 7
b 13 – 6 = 7 6 + 7 = 13
c 19 – 14 = 5 5 + 14 = 19

Page 5

2 **a** 14 + 5 = 19 19 – 5 = 14
b 12 + 3 = 15 15 – 3 = 12
c 9 + 8 = 17 17 – 8 = 9
d 11 + 9 = 20 20 – 11 = 9

3

10 – 4	19 – 8	8 + 9	16 – 3	6 + 5
11 + 8	11 – 5	13 + 3	17 – 8	6 + 4

Page 6

1 **a** 9 + 4 + 2 = 15
b 8 + 5 + 4 = 17
c 7 + 5 + 0 = 12

Page 7

2 **b** 2 + 6 + 8 = 8 + 2 + 6 = 10 + 6 = 16
c 3 + 5 + 7 = 7 + 3 + 5 = 10 + 5 = 15

3 **a** 3 + 5 + 3 = 11
b 2 + 9 + 2 = 13
c 5 + 7 + 7 = 19
d 6 + 6 + 1 =13

Fun Zone

Example words: stem, no, rent, net, met, rest, on, stone

Page 8

1 **a** 28, 27, 26, 25, 24, 23
b 69, 70, 71, 72, 73, 74
c 57, 58, 59, 60, 61, 62
d 22, 21, 20, 19, 18, 17
e 87, 88, 89, 90, 91, 92

Page 9

2 **b** 76 + 8 = 84 **d** 88 – 9 = 79
c 39 – 7 = 32

3 **a** 37 + 8 = 45 **c** 87 + 3 = 90
b 23 + 9 = 32 **d** 57 – 7 = 50

Fun Zone

Start
Finish

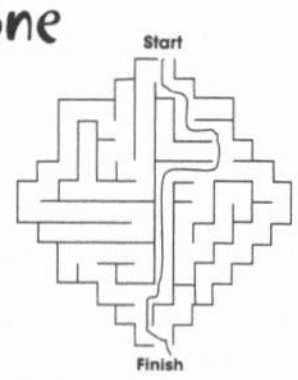

Page 10

1 **a** 4 + 3 = 7 40 + 30 = 70
b 9 + 4 = 13 90 + 40 = 130
c 6 + 2 = 8 60 + 20 = 80
d 5 + 3 = 8 50 + 30 = 80
e 7 + 5 = 12 70 + 50 = 120
f 8 + 7 = 15 80 + 70 = 150

Page 11

2 **a** 30 + 50 = 80
b 70 + 20 = 90
c 60 + 50 = 110
d 90 + 40 = 130
e 40 + 40 = 80
f 40 + 60 = 100
g 80 + 60 = 140
h 70 + 90 = 160

3

90 + 70
70 + 60
80 + 70
130
90
150
140
110
160
40 + 50
30 + 80
60 + 80

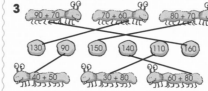

Page 12

1 **b** 76 + 10 = 86
c 37 + 10 = 47
d 43 + 10 = 53

2 **b** 55 + 30 = 85
c 40 + 40 = 80
d 12 + 40 = 52

Page 13

3 **b** 17 – 10 = 7
c 44 – 10 = 34
d 36 – 10 = 26

4 **b** 99 – 40 = 59
c 58 – 30 = 28
d 61 – 20 = 41

5

4 + 30
26 + 40
52 – 20
7
34
19
32
66
100
17 – 10
80 + 20
39 – 20

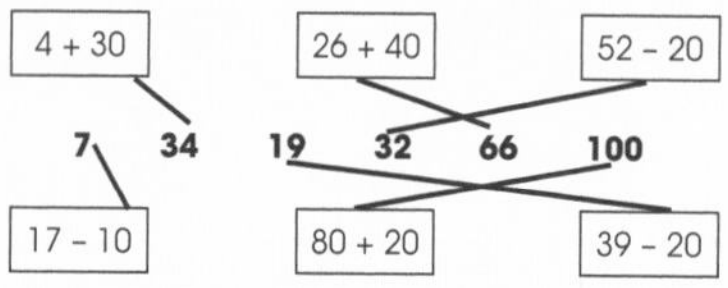

Page 14

1 **a**

+	8	7	6
4	12	11	10
7	15	14	13
9	17	16	15

b

+	13	12	8
7	20	19	15
6	19	18	14
5	18	17	13

2 **a** 5 + 6 = 11; 6 + 5 = 11; 11 – 6 = 5; 11 – 5 = 6
b 14 – 10 = 4; 14 – 4 = 10; 10 + 4 = 14; 4 + 10 = 14
c 5 + 2 = 7; 2 + 5 = 7; 7 – 2 = 5; 7 – 5 =
d 9 – 4 = 5; 9 – 5 = 4; 5 + 4 = 9; 4 + 5 =

3 **b-d** Child to add two small jumps on the number line and complete number sentence.

Page 15

4 **a** 14 – 7 = 7; 23 – 7 = 16; 44 – 7 = 37; 67 – 7 = 60
b 8 + 8 = 16; 41 + 8 = 49; 56 + 8 = 64; 77 + 8 = 85

5 **a** 90 – 30 = 60
b 100 – 60 = 40
c 70 + 10 = 80
d 120 – 10 = 110
e 50 – 40 = 10
f 60 – 50 = 10

6

42 + 30
96 – 20
146 – 40
106
84
93
72
76
105
155 – 50
33 + 60
34 + 50

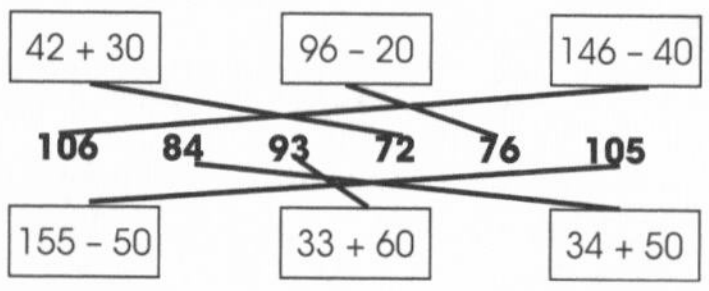

Page 16

1 **b** 23 = 20 + 3 **e** 70 = 70 + 0
c 98 = 90 + 8 **f** 39 = 30 + 9
d 15 = 10 + 5

Page 17

2 **a** 27 + 31 = 20 + 30 + 7 + 1 = 50 + 7 + 1 = 58
b 42 + 34 = 40 + 30 + 2 + 4 = 70 + 2 + 4 = 76

3

41 + 15
31 + 34
24 + 32
21 + 52
56
65
73
42 + 31
21 + 44
29 + 27

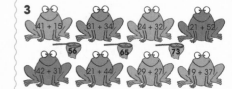

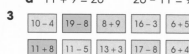

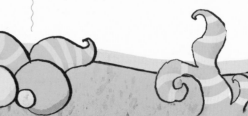

Fun Zone

Page 18

1 a 48 – 16 = 32

  b 56 – 13 = 43

  c 75 – 14 = 61

Page 19

2 a 0, 15, 30, 45, 60, 75, 90

  b 0, 11, 22, 33, 44, 55, 66, 77

3 a 43 – 21 = 22   54 – 32 = 22

    62 – 40 = 22   36 – 15 = 21

  b 74 – 21 = 53   85 – 33 = 52

    96 – 43 = 53   87 – 34 = 53

  c 56 – 12 = 44   65 – 31 = 34

    77 – 43 = 34   58 – 24 = 34

Page 20

1 a 24 – 12 = 12

  b 35 – 17 = 18

  c 52 – 14 = 38

Page 21

2 25 + 35 = 60; 37 + 40 = 77;
14 + 9 = 23; 87 + 7 = 94;
9 + 18 = 27; 42 + 13 = 55;
17 + 14 = 33

3

a 7 + 8 = 15   **d** 30 – 17 = 13   **g** 5 + 12 = 17

b 98 – 18 = 80   **e** 16 – 8 = 8   **h** 7 + 9 = 16

c 17 + 15 = 32   **f** 76 – 58 = 18   **i** 55 – 15 = 40

Fun Zone

50	12	13	14	15	16	42
20	51	4	34	87	22	41
82	18	52	65	11	35	40
83	44	1	53	67	94	39
84	29	88	3	54	32	38
85	72	71	70	69	68	24
86	99	6	5	7	8	9

Page 22

1

	a t	h	r	e	e	
b e	i	g	h	t		
	c s	i	x			
d f	o	u	r			
	e t	w	e	l	v	e
	f s	e	v	e	n	
g n	i	n	e			
	h o	n	e			

Page 23

2 a Five < ten

  b Eleven > seven

---

  c Ninety-nine > thirty-six

  d Fifty-one > twelve

3 a 30 – 16 = 14

  b 30 + 28 = 58

  c 30 – 11 = 19

Fun Zone

Q	G	R	U	N	T	L
C	S	B	M	M	I	W
R	E	N	O	A	J	O
Q	O	R	A	F	Y	R
V	A	A	N	R	E	G
T	O	X	R	C	L	U
P	H	I	S	S	L	W

Page 24

1 a

4	3			
+ 2	6			
	9	(	3 +	6)
6	0	(	40 +	20)
6	9			

c

6	7			
+ 3	1			
	8	(	7 +	1)
9	0	(	60 +	30)
9	8			

b

5	1			
+ 3	6			
	7	(	1 +	6)
8	0	(	50 +	30)
8	7			

d

6	4			
+ 2	5			
	9	(	4 +	5)
8	0	(	60 +	20)
8	9			

Page 25

2 42 + 23 = 65; 30 + 23 = 53;
25 + 23 = 48; 66 + 23 = 89

3 a 79 cm **c** 99 cm **e** 31 cm

  b 79 cm **d** 49 cm

Page 26

1 b

4 0	+	7
– 3 0	+	6
1 0	+	1 = 1 1

c

9 0	+	8
– 2 0	+	3
7 0	+	5 = 7 5

Page 27

2 77 – 32 = 45; 65 – 32 = 33;
89 – 32 = 57; 43 – 32 = 11

3 a

5 0	+	7
– 2 0	+	1
3 0	+	6 = 3 6

b

4 0	+	5
– 3 0	+	0
1 0	+	5 = 1 5

---

c

5 0	+	8
– 4 0	+	5
1 0	+	3 = 1 3

Fun Zone

Example poem: **M**uddy

    Ogre

    Naughty

    Scary

    Teeth

    Enormous

    Roar

Page 28

1 a 52 + 41 = 50 + 40 + 2 + 1 =
90 + 2 + 1 = 93

  b 24 + 73 = 20 + 70 + 4 + 3 =
90 + 4 + 3 = 97

2 a 37 **b** 79 **c** 7 **d** 69

3 a 27 **b** 19 **c** 38 **d** 41

Page 29

4

owt	two	60
veleen	eleven	8
hirtty	thirty	2
tixys	sixty	30
higet	eight	11

5 a

3	6			
+ 2	3			
	9	(	6 +	3)
5	0	(	30 +	20)
5	9			

b

5	7			
+ 4	1			
	8	(	7 +	1)
9	0	(	50 +	40)
9	8			

6 a 15p **b** 20p **c** 24p **d** 34p

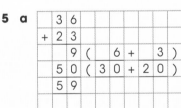
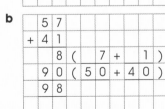

Monster Match

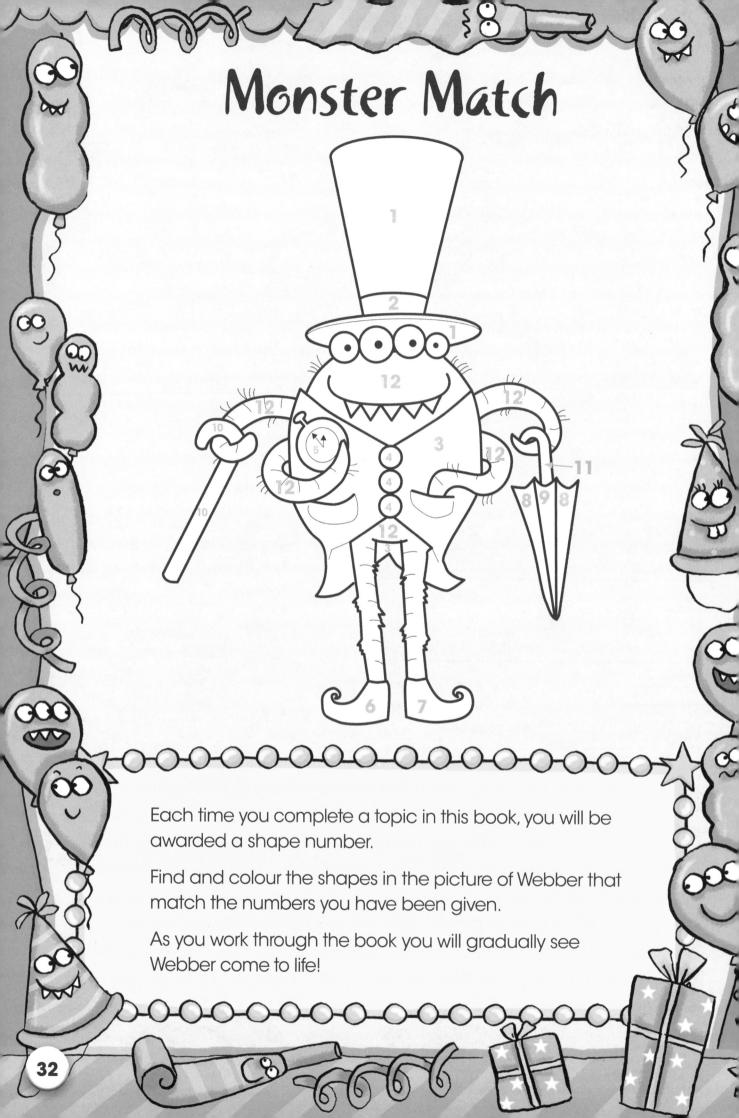

Each time you complete a topic in this book, you will be awarded a shape number.

Find and colour the shapes in the picture of Webber that match the numbers you have been given.

As you work through the book you will gradually see Webber come to life!